Benjamin Sullivan

Contents

Introduction

This publication is comprised of two equations which are derived and extrapolated from a book that I created in 2015/2016: Probability, Quantum Mechanics, and Probability-Quanta. The latter consists of various thought experiments and postulates. I have included scans from the original book in an effort to demonstrate my work and in the hopes of having the experiments conducted.

I do not have any formal education in physics and I realize that producing the publication without any academic background will see the work be dismissed or seen as suspect by many.

What I do possess is the ability to know when something is right.

Something is right.

Title: Sullivan Unifying Constant
Author: Benjamin Allen Sullivan
Comments: Equation, 2 pages with 9 Figures
Subj-class: Physics

The Sullivan Unifying Constant is an equation that resolves the long-sought pursuit of understanding the unification of the known forces in the universe and how they interact with each other. This equation can be applied to or within any of the known sciences. The equation incorporates and demonstrates the interaction between forces known within "Classical Physics" and merges them with modern "Quantum Theory".

1. Each individual component of the equation has been professionally pursued with vigour by various people, schools, and various scientific organizations. As mentioned, the equation incorporates and demonstrates the interaction between forces known within "Classical Physics" and merges them with modern "Quantum Theory". It accomplishes this by incorporating the Spin Values (SV) of the Masses.

2. My equation resolves the long-sought pursuit of understanding the unification of the known forces in the universe and how they interact with each other. It does this by demonstrating the interchangeable nature of the known Force Carriers within the equation:

3. Figures:

Figure 1: Sullivan Unifying Constant Equation

$$FC = T \frac{FCm1\mathbf{\uparrow}^{sv} FCm2\mathbf{\downarrow}^{sv}}{FCd^2}$$

Figure 2 FC = Any Force Carrier

$$FC$$

FC = G for Gravity, EM for Electromagnetism, SF for Strong Force, WF for Weak Force.

Figure 3: Force Carriers

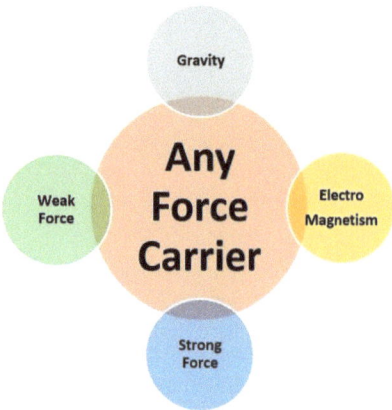

Figure 4: Sullivan Unifying Constant Equation Example Using Force of Gravity:

$$F = \text{T} \frac{\text{SF}m1\Uparrow^{sv}\text{WF}m2\Downarrow^{sv}}{\text{EM}d^2}$$

Figure 5: Equation Component #1 - Force of Gravity.

$$F$$

Figure 6: Equation Component #2 – Time (Constant).

Figure 7: Equation Component #3 - Strong Force (SF) - Mass 1 (m1) with indication of Spin Value (SV) Entangled or Otherwise.

$$SFm1\uparrow^{SV}$$

Figure 8: Equation Component #4 - Weak Force (WF) - Mass 2 (m2) with indication of Spin Value (SV) Entangled or Otherwise.

$$WFm2\downarrow^{SV}$$

Figure 9: Equation Component #5 – Electro Magnetism (EM) and Distance (d2).

$$EMd^2$$

4. Description: The Sullivan Unifying Constant demonstrates the interaction of the known forces within universe.

5. Examples of use: The Sullivan Unifying Constant provides enlightenment within all sciences. The equation offers a framework of understanding that allows for the creation of new technologies by progressing the field of science forward with new vigour derived from comprehension of the equation. Many of the implications (technological and otherwise) of the equation have yet to be discovered due to its newly formalized conception. It requires the contemplation of all with interest to achieve its maximum potential.

Title: Sullivan-Newtonian Gravitational Equation
Author: Benjamin Allen Sullivan
Comments: Equation, 2 pages with 6 Figures
Subj-class: Physics

The Sullivan-Newtonian Gravitational Equation is related to Gravity within the field of Physics and all other affected Sciences. The equation is an altercation of Sir Isaac Newton's Universal Law of Gravitation. The equation merges and demonstrates the interaction between what is known as "Classical Physics" and modern "Quantum Theory".

1. As mentioned, the equation merges and demonstrates the interaction between what is known as "Classical Physics" and modern "Quantum Theory". It accomplishes this by incorporating the Spin Values (SV) of the Masses into the original equation. Individual components of the original Sir Isaac Newton equation have been professionally pursued with vigour by various people, schools, and various scientific organizations.

2. My equation resolves the long-sought pursuit of understanding the complete nature of Gravity.

3. Figures:

Figure 10: Sullivan-Newtonian Gravitational Equation

$$F = G \frac{M1\!\uparrow^{SV} \; M2\!\downarrow^{SV}}{D^2}$$

Figure 11: Equation Component #1 - Force of Gravity

$$F$$

Figure 12: Equation Component #2 – Constant

$$G$$

Figure 13: Equation Component #3 - Mass 1 (M1) with indication of Spin Value (SV) Entangled or Otherwise.

$$M1 \uparrow^{SV}$$

Figure 14: Equation Component #4 - Mass 2 (M2) with indication of Spin Value (SV) Entangled or Otherwise.

$$M2 \downarrow^{SV}$$

Figure 15: Equation Component #5 – Distance (D2)

$$D^2$$

1. Description: The Sullivan-Newtonian Gravitational Equation demonstrates a comprehensive framework of Gravity.

5. Examples of use: The Sullivan-Newtonian Gravitational Equation provides enlightenment within all sciences affected by Gravity. The equation offers a framework of understanding that allows for the creation of new technologies by progressing the field of science forward with new vigour derived from comprehension of the equation. Many of the implications (technological and otherwise) of the equation have yet to be discovered due to its newly formalized conception. It requires the contemplation of all with interest to achieve its maximum potential.

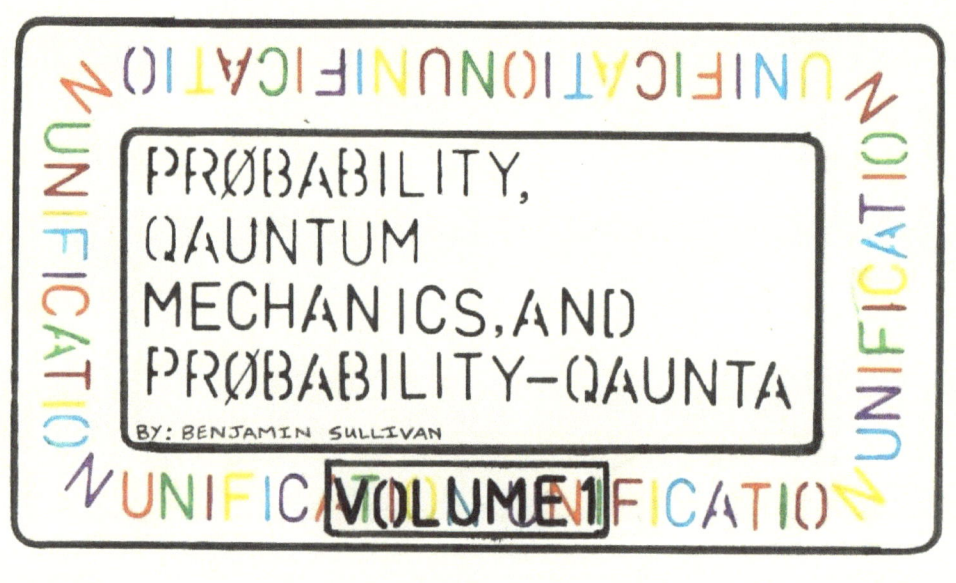

UNIFICATION UNIFICATION UNIFICATION

PRØBABILITY,
QAUNTUM
MECHANICS, AND
PRØBABILITY-QAUNTA

BY: BENJAMIN SULLIVAN

VOLUME 1

FOREWARD

PROBABILITY, QANUNTAM MECHANICS, AND PROBABILITY - QANUNTA
UNIFICATION WAS WRITTEN BY MYSELF OVER A PERIOD OF

ALTERED FORMULA

$$F = G \frac{M_1 \; M_2}{D^2}$$

NEWTONS LAW OF GRAVITY

CLASSIC

PROPOSED ALTERED FORMULA
(ENTANGLED PARTICLES / MASS SPIN)

PROPOSED

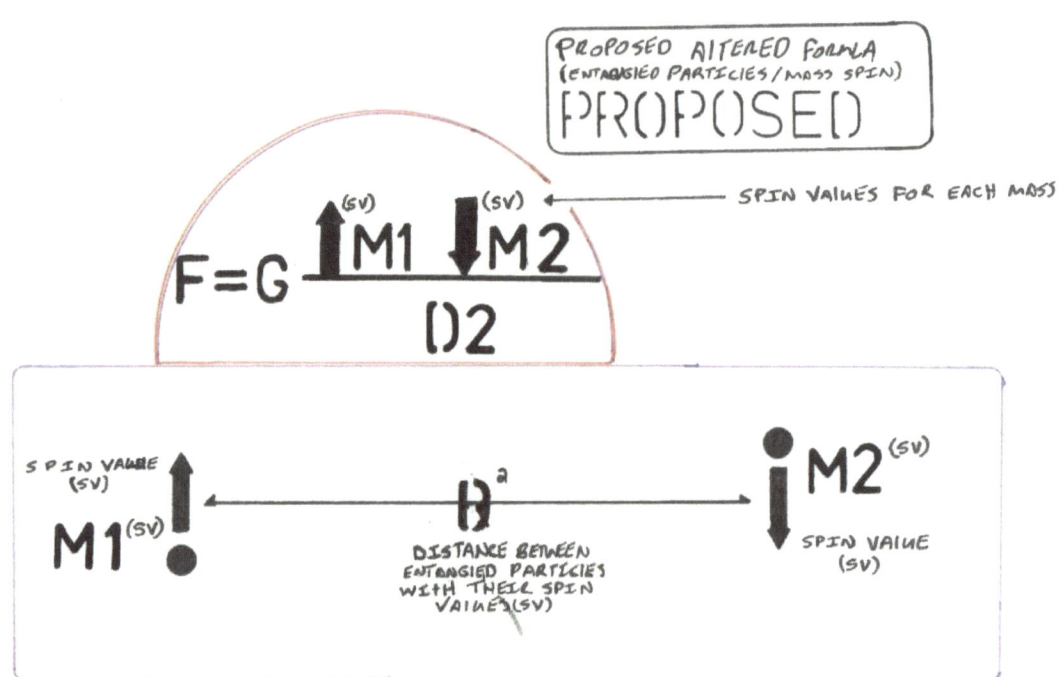

SPIN VALUES FOR EACH MASS

$$F = G \frac{\uparrow^{(sv)} M_1 \; \downarrow^{(sv)} M_2}{D^2}$$

SPIN VALUE (SV)

M1 $^{(sv)}$ ↑

D^2

DISTANCE BETWEEN ENTANGLED PARTICLES WITH THEIR SPIN VALUES (SV)

M2 $^{(sv)}$ ↓

SPIN VALUE (SV)

PROPOSED EXPERIMENTS:

- VARIOUS PARTICLES WITH DIFFERENT WEIGHTS / MASS / SPIN(S)
- LOW KELVIN EXPERIMENT WITH THE SAME DATA
- VARIATIONS OF THE EXPERIMENT WITH DIFFERENT SPIN VALUES (SV) AND MASSES
- VARIOUS ENVIRONMENTS OF ELECTROMAGNETISM

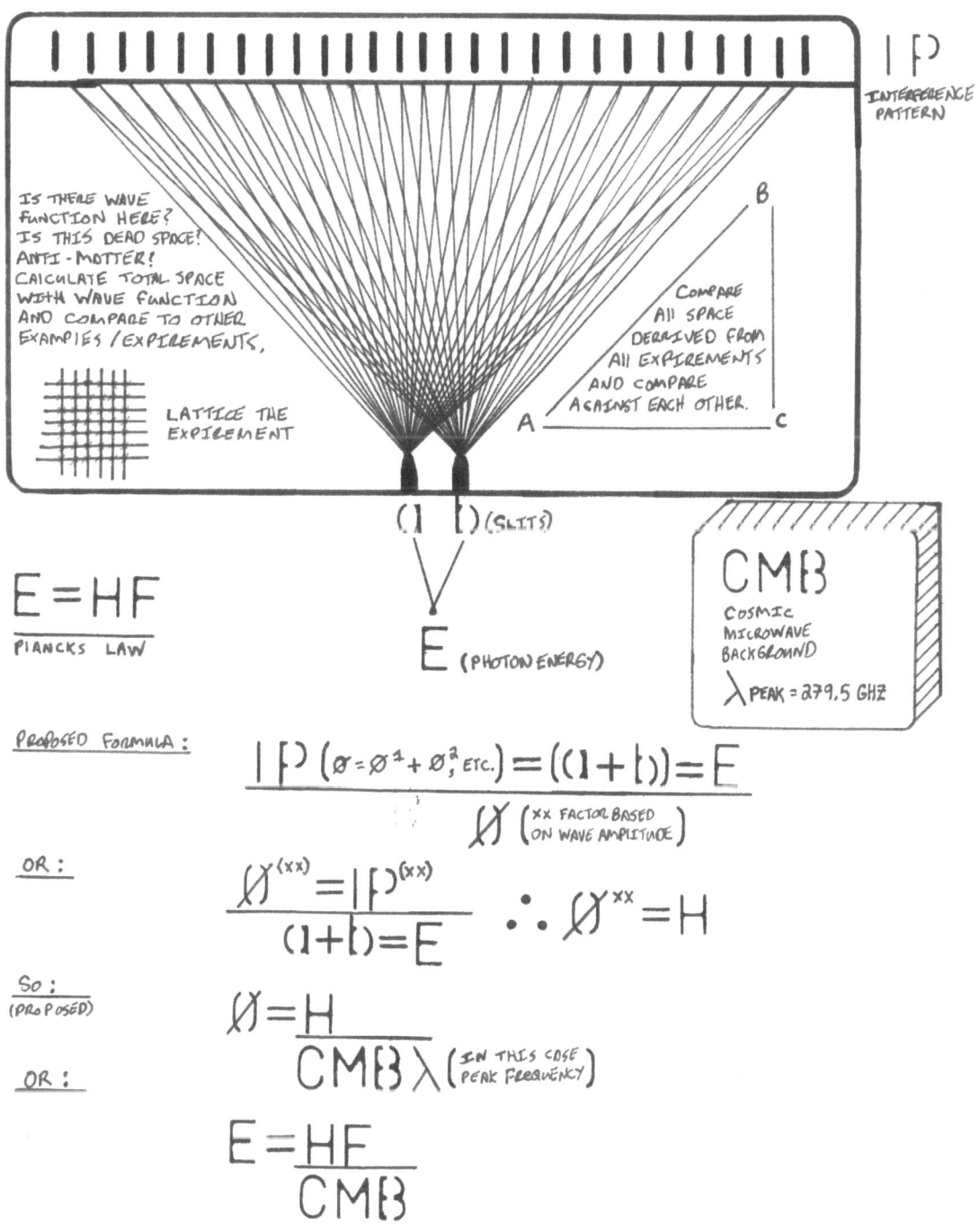

IP
INTERFERENCE PATTERN

IS THERE WAVE FUNCTION HERE?
IS THIS DEAD SPACE?
ANTI-MATTER!
CALCULATE TOTAL SPACE WITH WAVE FUNCTION AND COMPARE TO OTHER EXAMPLES / EXPIREMENTS.

LATTICE THE EXPIREMENT

COMPARE ALL SPACE DERRIVED FROM All EXPIREMENTS AND COMPARE AGAINST EACH OTHER.

(| |)(SLITS)

E (PHOTON ENERGY)

$$E = HF$$
PLANCKS LAW

CMB
COSMIC MICROWAVE BACKGROUND
λ PEAK = 279.5 GHZ

PROPOSED FORMULA:

$$\frac{IP(\emptyset = \emptyset^2 + \emptyset_3^2 \text{ ETC.}) = ((1+b)) = E}{\emptyset} \text{(xx FACTOR BASED ON WAVE AMPLITUDE)}$$

OR:

$$\frac{\emptyset^{(xx)} = IP^{(xx)}}{(1+b) = E} \quad \therefore \emptyset^{xx} = H$$

SO: (PROPOSED)

$$\emptyset = \frac{H}{CMB\lambda} \text{(IN THIS CASE PEAK FREQUENCY)}$$

OR:

$$E = \frac{HF}{CMB}$$

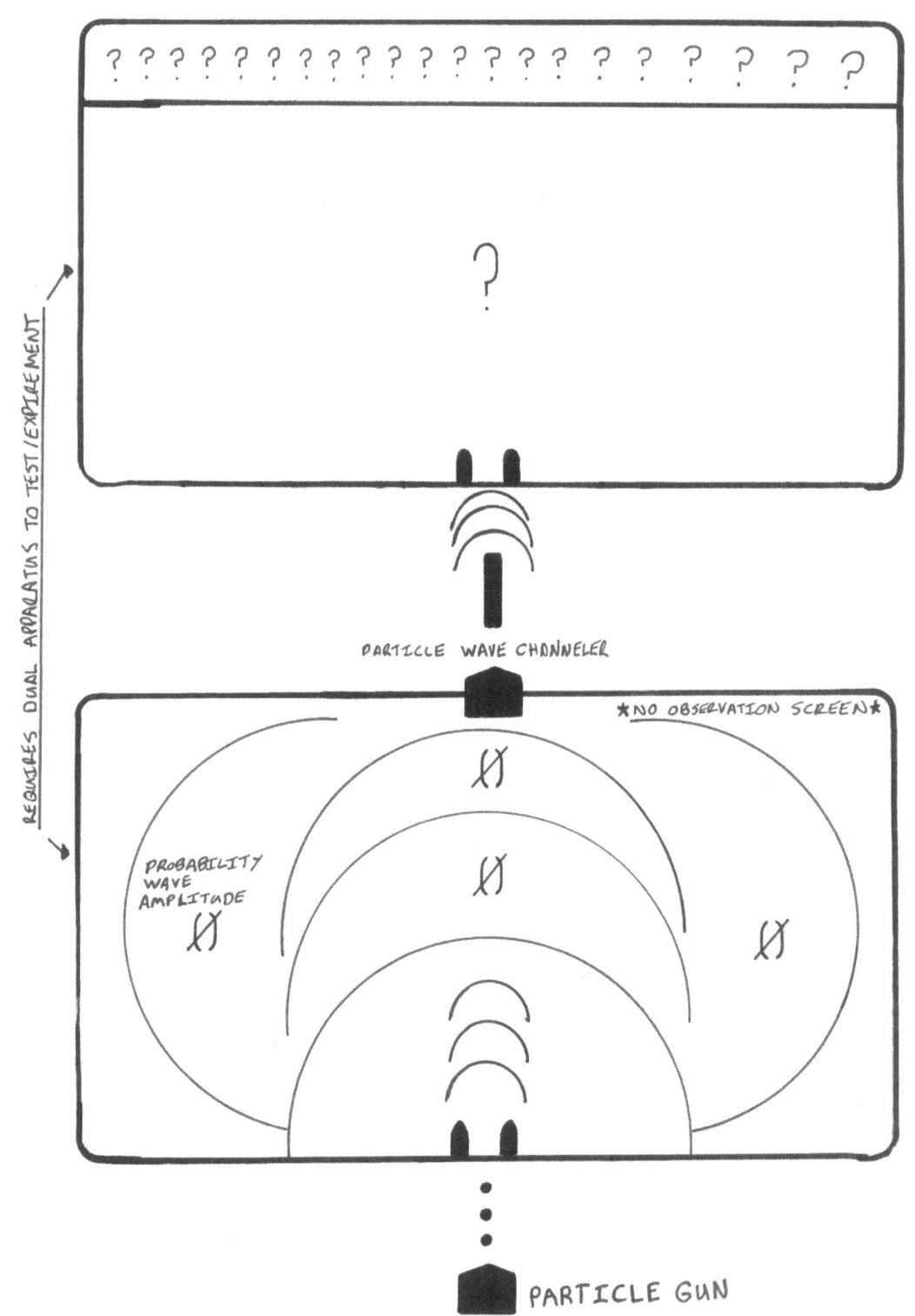

? ? ? ? ? ? ? ? ? ? ? ? ? ? ? ?

?

REQUIRES DUAL APPARATUS TO TEST / ENTICEMENT

PARTICLE WAVE CHANNELER

NO OBSERVATION SCREEN

PROBABILITY WAVE AMPLITUDE

PARTICLE GUN

PROPOSED REVERSE EXPIREMENT-USE INTERFERENCE
PATTERN DATA "FIRST" AND REVERSING THE DATA
THROUGH TWO SLITS:

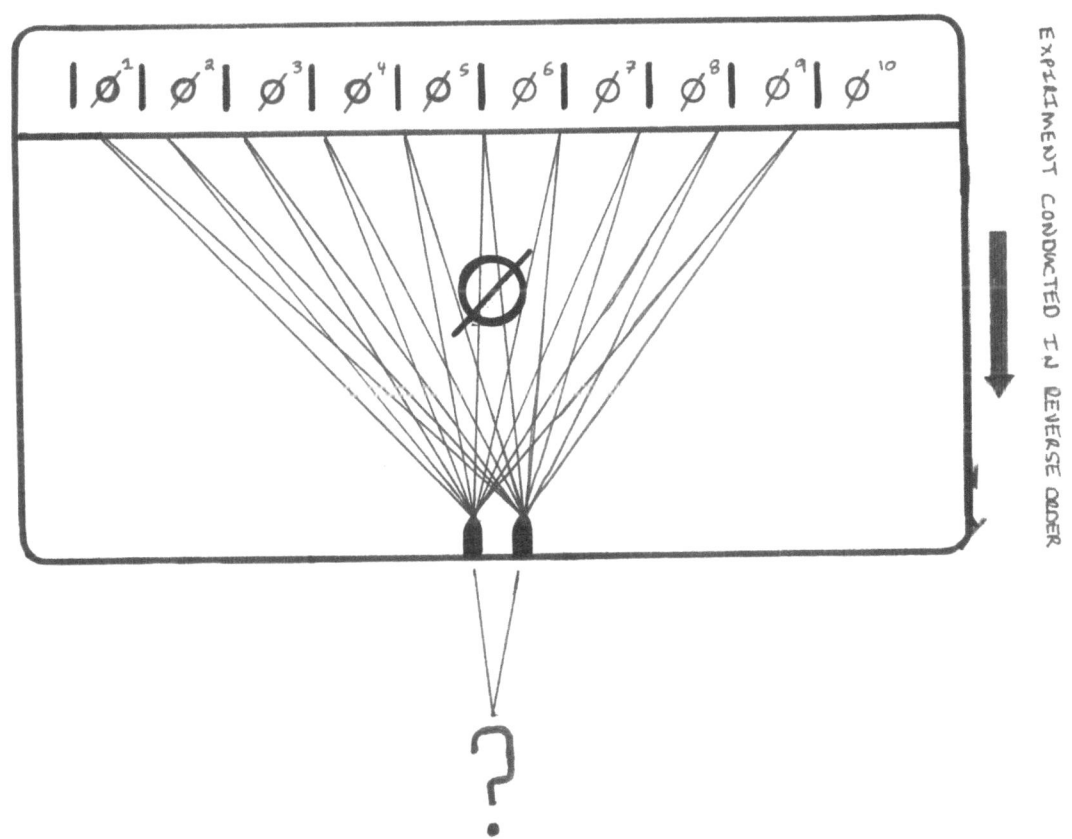

PROPOSED ENTANGLED PARTICLES ONLY EXPIRIMENT

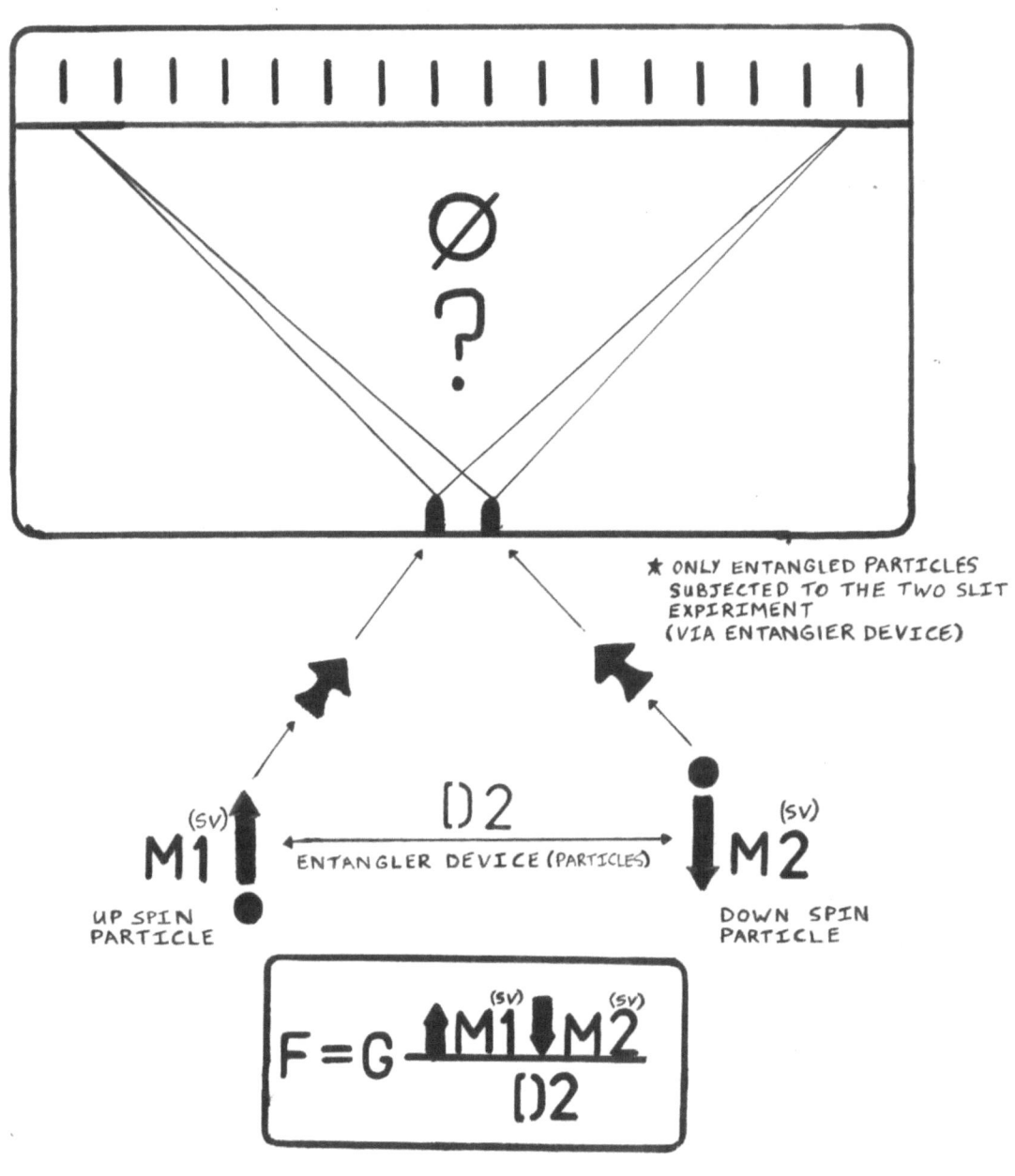

★ ONLY ENTANGLED PARTICLES
SUBJECTED TO THE TWO SLIT
EXPIRIMENT
(VIA ENTANGIER DEVICE)

D2
ENTANGLER DEVICE (PARTICLES)

M1 (sv)   M2 (sv)

UP SPIN
PARTICLE

DOWN SPIN
PARTICLE

$$F = G \frac{\uparrow M1^{(sv)} \downarrow M2^{(sv)}}{D2}$$

CONCEPTUAL VISUAL UNIFICATION AID

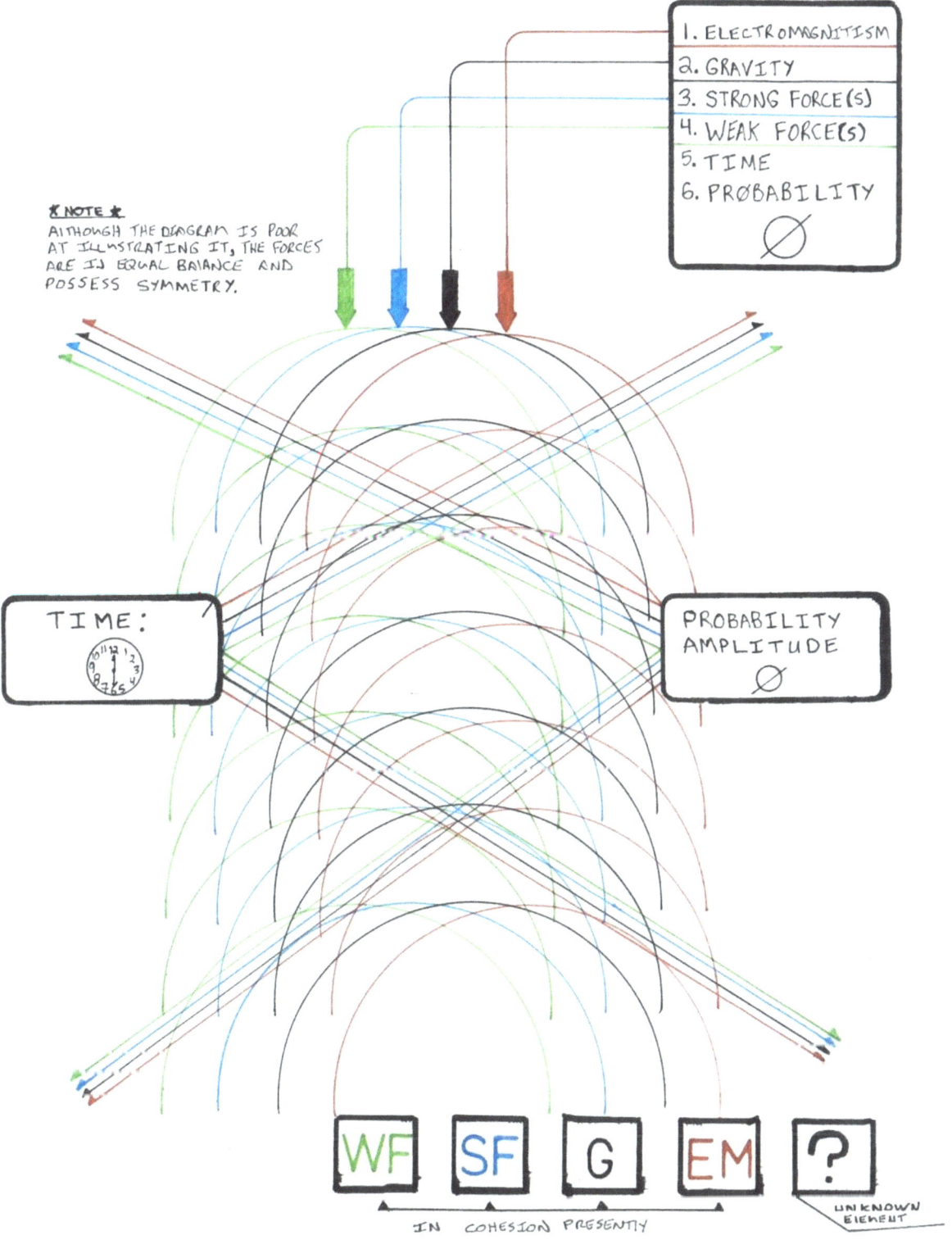

1. ELECTROMAGNITISM
2. GRAVITY
3. STRONG FORCE(S)
4. WEAK FORCE(S)
5. TIME
6. PRØBABILITY
Ø

★ NOTE ★
ALTHOUGH THE DIAGRAM IS POOR
AT ILLUSTRATING IT, THE FORCES
ARE IN EQUAL BALANCE AND
POSSESS SYMMETRY.

TIME:

PROBABILITY
AMPLITUDE
Ø

WF SF G EM ?

IN COHESION PRESENTLY

UNKNOWN
ELEMENT

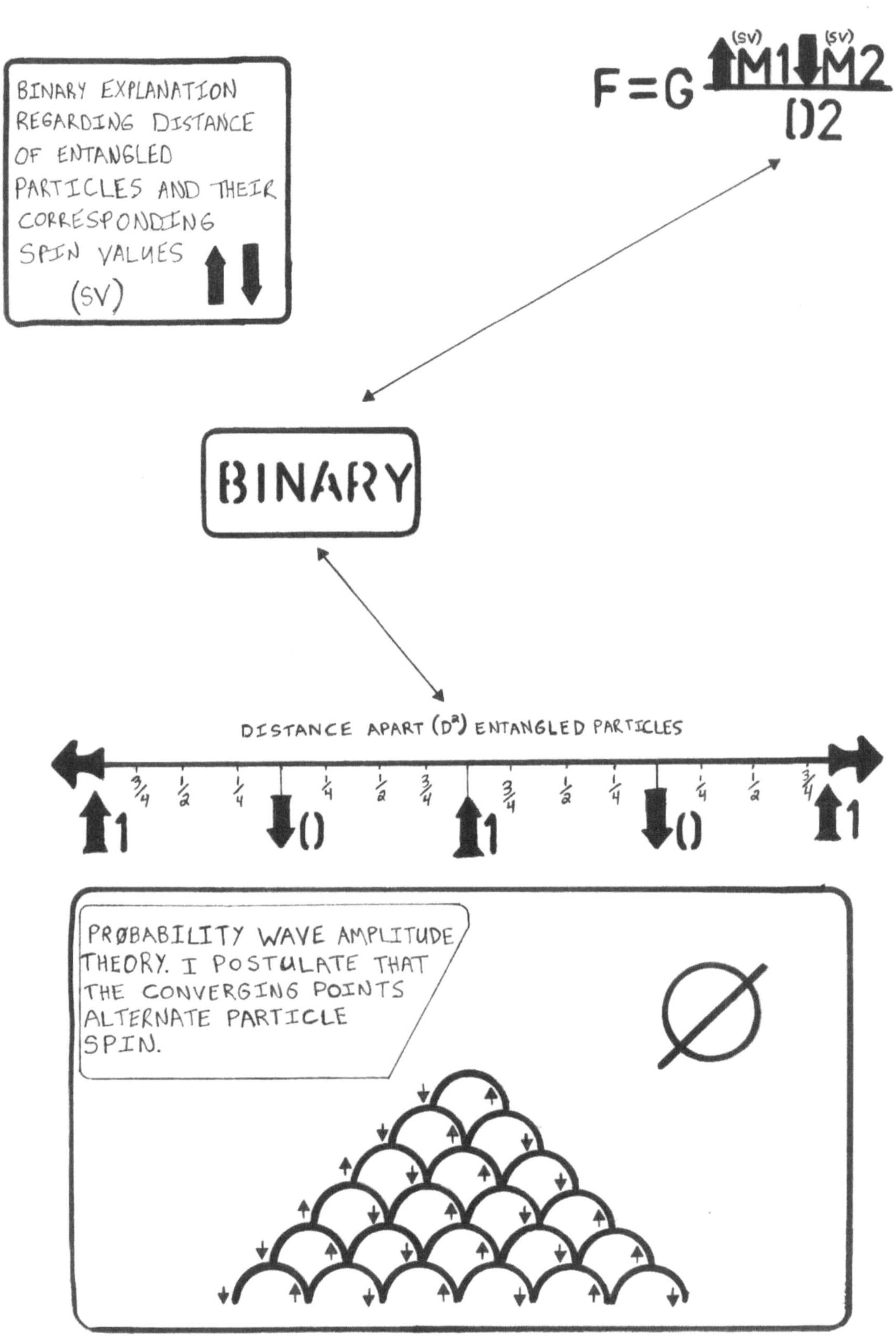

BINARY EXPLANATION REGARDING DISTANCE OF ENTANGLED PARTICLES AND THEIR CORRESPONDING SPIN VALUES (SV) ⬆⬇

$$F = G \frac{\overset{(sv)}{\Uparrow} M1 \overset{(sv)}{\Downarrow} M2}{D2}$$

BINARY

DISTANCE APART (D²) ENTANGLED PARTICLES

⬆1   ¾ ½ ¼   ⬇()   ¼ ½ ¾   ⬆1   ¾ ½ ¼   ⬇()   ¼ ½ ¾   ⬆1

PROBABILITY WAVE AMPLITUDE THEORY. I POSTULATE THAT THE CONVERGING POINTS ALTERNATE PARTICLE SPIN.

⊘

ENTANGLED PARTICLE SPIN VALUE (SV)
DETERMINED BY WAVE AMPLITUDE ∅

BINARY SYSTEM
WAVE CONVERGING POINTS

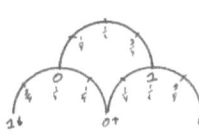

D2

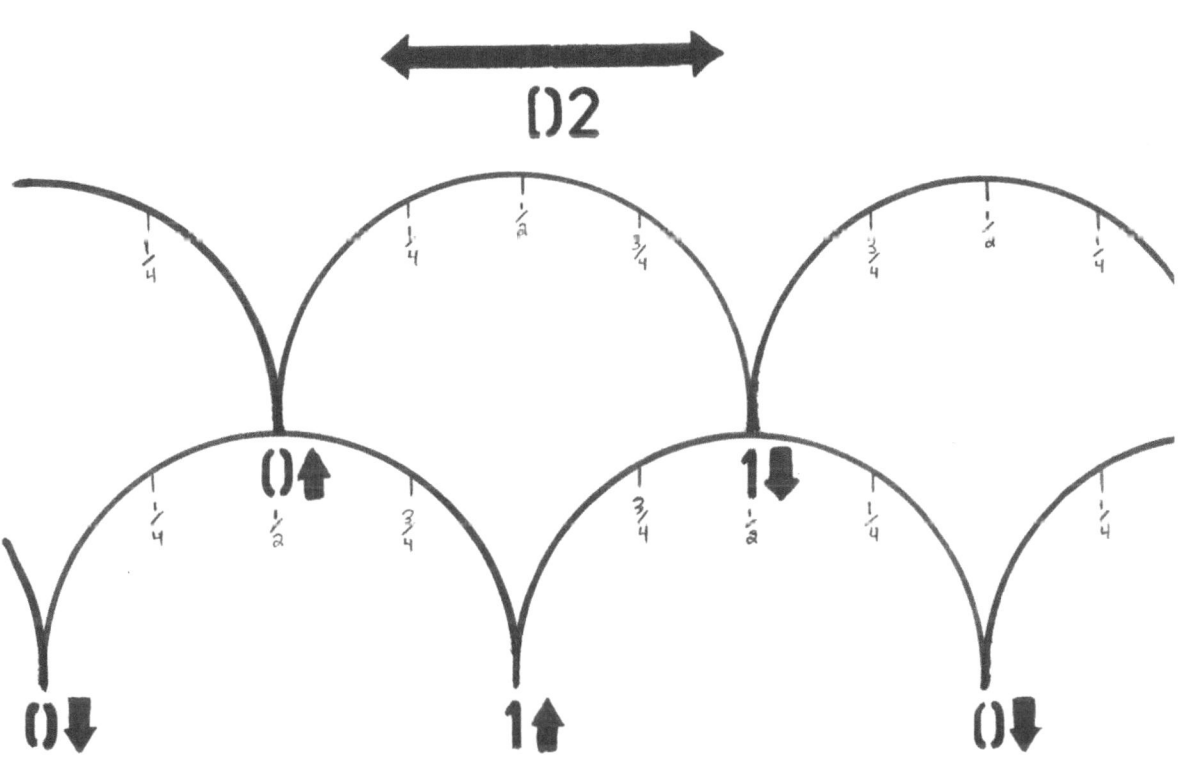

$$F = G \frac{\uparrow M1^{(sv)} \downarrow M2^{(sv)}}{D2}$$

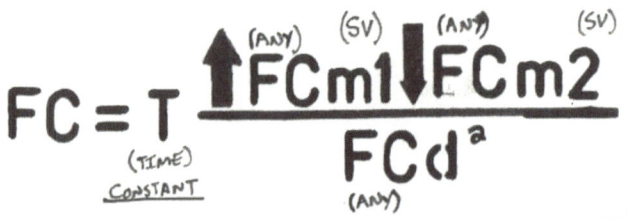

$$FC = T \frac{\uparrow FCm1 \downarrow FCm2}{FCd^a}$$

(ANY) (SV) (ANY) (SV)

(TIME) CONSTANT

(ANY)

FC = ANY FORCE CARRIER ⟶

- GRAVITY
- ELECTROMAGNETISM
- STRONG FORCE
- WEAK FORCE

G
WF   EM
SF

UNIFYING CONSTANT

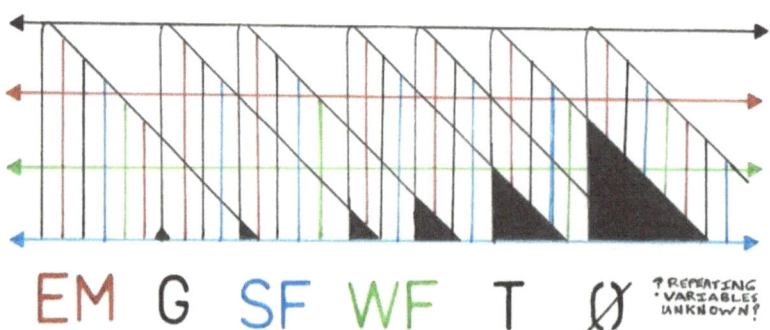

EM   G   SF   WF   T   Ø   ? REPEATING
VARIABLES
UNKNOWN?

PROPOSED UNIFYING
THEORY
(GRAVITY EXAMPLE)

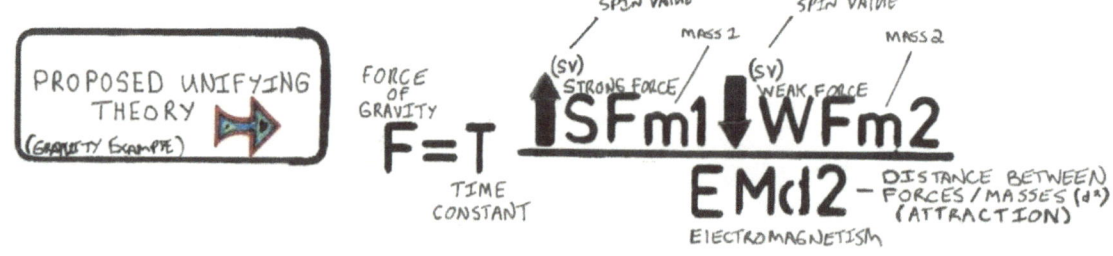

SPIN VALUE   SPIN VALUE
(SV)   MASS 1   (SV)   MASS 2
STRONG FORCE   WEAK FORCE

$$F = T \frac{\uparrow SFm1 \downarrow WFm2}{EMd2}$$

FORCE OF GRAVITY

TIME CONSTANT

ELECTROMAGNETISM

— DISTANCE BETWEEN FORCES / MASSES (d²) (ATTRACTION)

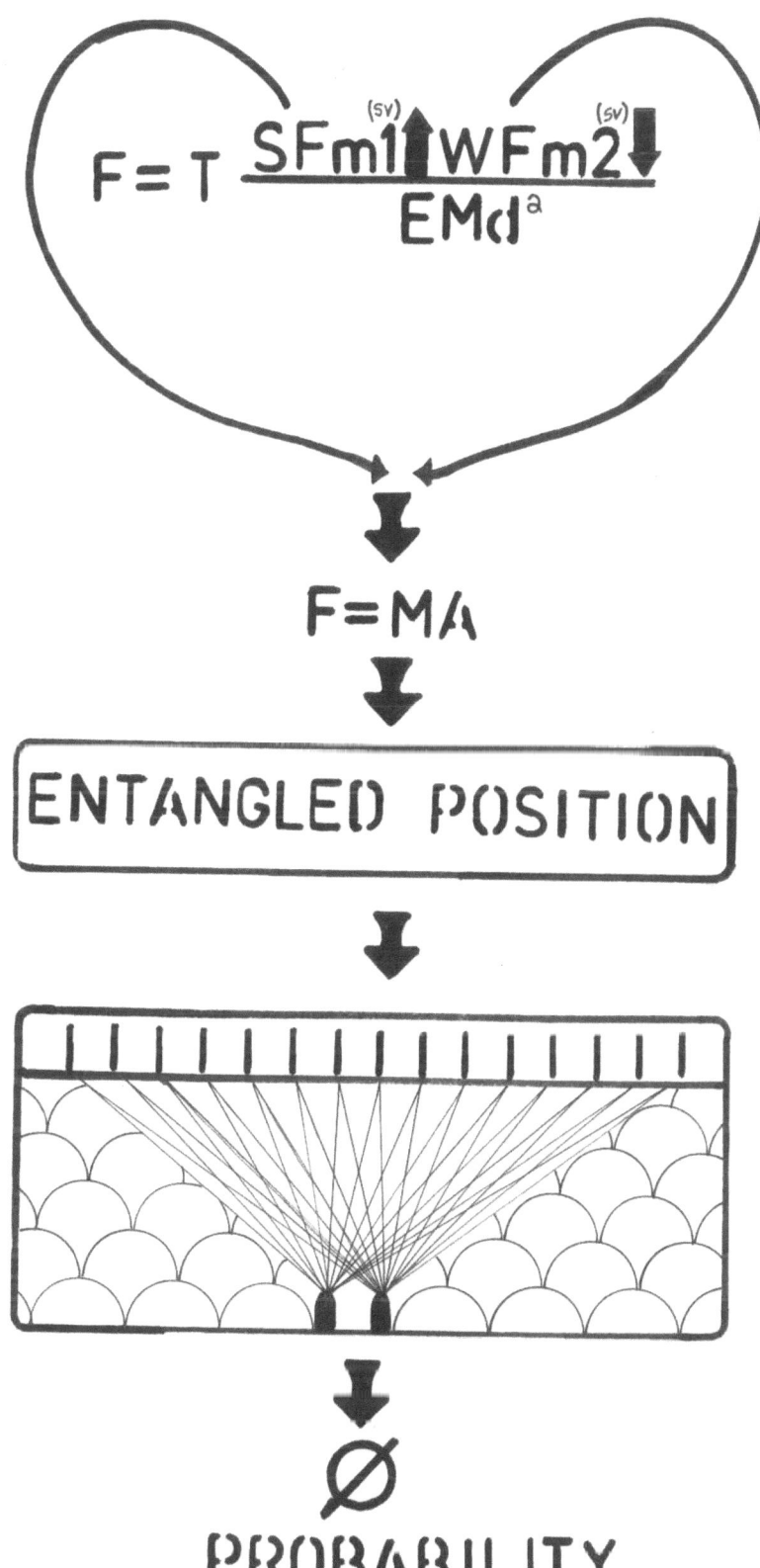

$$F = T \, \frac{SFm1\uparrow^{(sv)} WFm2\downarrow^{(sv)}}{EMd^a}$$

$F = MA$

ENTANGLED POSITION

⌀

PROBABILITY

$$F = G \frac{\uparrow^{(sv)} M1 \downarrow^{(sv)} M2}{D2}$$

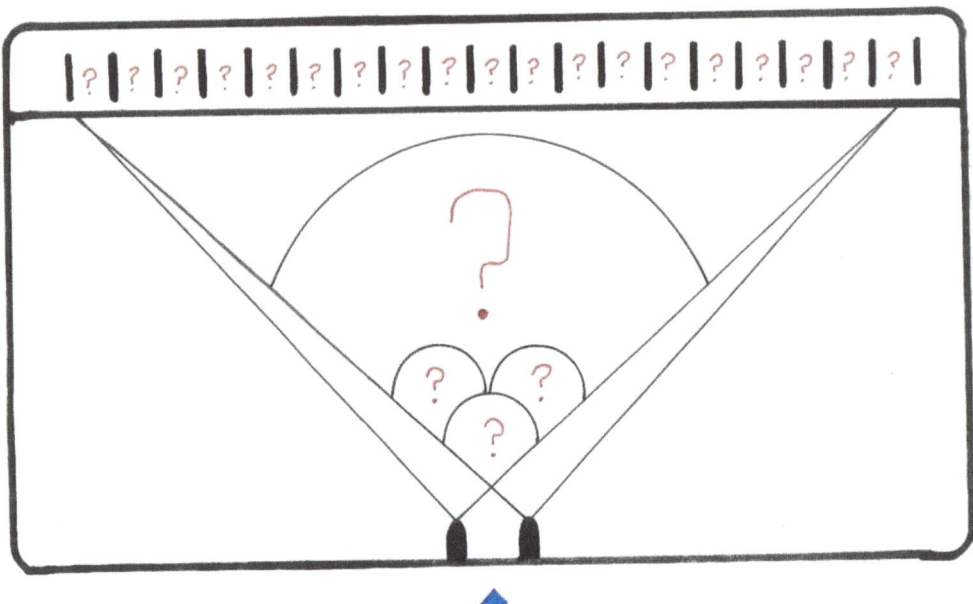

1. WHAT EFFECT DOES THE DISTANCE BETWEEN THE PARTICLE GUN AND OBSERVATION SCREEN HAVE ON THE EXPIRIMENT ???

2.

$$F = M/A^{(sv)}$$

FORCE = MASS$^{(sv)}$ X ACCELERATION WITH THE ADDITION OF THE SPIN VALUE (SV) ANNOTATED / MEASURED FOR THE ENTANGLED PARTICLE (S) ENTERING THE EXPIRIMENT.

ENTANGLED PARTICLE GUN (SV)^2

THE MAJORITY OF QM EXPIREMENTS HAVE BEEN CONDUCTED AT LOW TEMPRATURE CONDITIONS (WITH PROGRESSIVE EFFORT TOWARDS WARMER TEMPROTURES). I POSTULATE THAT WE ARE MISSING SIGNIFICANT ELEMENTS OF QM THEORY BY NOT CONDUCTING EXPIRIMENTS AT HIGH TEMPROTURES. WITH THE REALIZATION THAT HIGH TEMPRATURES WITHIN QM EXPERIMENTS ARE DIFFICULT TO DETECT, EVERY EFFORT SHOULD BE MADE TO EXPIRIMENT AT DIII TEMPRATURES

HIGH-TEMP

MID

ULTRA-LOW

ALTERING SPIN VALUES (SV) OF ENTANGLED PARTICLES
FOR INFORMATION ENCODING AND TRANSMISSION

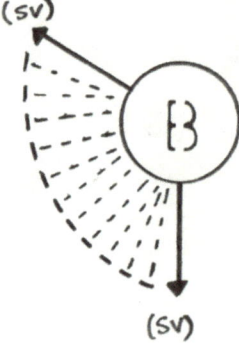

(SV)

(SV)

ELECTROMAGNETISM (EM) ADJUSTS ONE OF THE ENTANGLED PARTICLE'S SPIN VALUE (SV): IN THIS CASE BOB'S (B). THIS SIMULTANEOUSLY ADJUSTS THE SPIN VALUE OF IT'S ENTANGLED PARTNER: ALICE (A). NEITHER OF THE ENTANGLED PARTICLES ARE OBSERVED BEFORE, DURING, OR AFTER THE ADJUSTMENT OF THE SPIN VALUES.

D2 (ANY DISTANCE APART)

ALTERING THE SPIN VALUES IN THIS MANNER ALLOWS FOR THE ENCODING OF VARIOUS INFORMATION TYPES THAT MAY WORK BETTER IN ONE SPIN VALUE (SV) OPPOSED TO ANOTHER. THIS PROCESS WILL ALLOW FOR THE CHANGING OF ENTANGLED STATES SIMILAR TO THAT OF A TRADITIONAL "CRYPTO CHANGE". THIS WOULD BECOME RELEVANT WITH THE ADVENT OF ENTANGLED PARTICLE DETECTORS OR IF ENTANGLED "JAMMERS" WERE TO COME TO FRUITION.

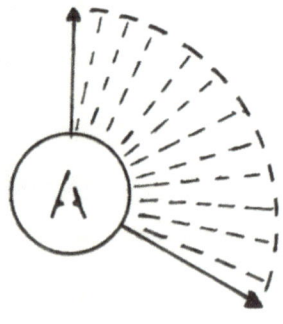

Conclusion

I welcome any correspondence on the content of the publication and I look forward to receiving your feedback.

Questions?

Postulates?

Problems?

Alterations?

Experimental Results?

Contact Information

Benjamin Allen Sullivan

Phone Number: 506-999-4934

E-Mail: nebyllus@yahoo.ca

Thoughts